Molecular Docking: A Formula Handbook

N.B. Singh

DEDICATION

To Nature,

I dedicate this book to you, the source of all life. You are my inspiration, my teacher, and my friend.

Thank you for teaching me about the beauty of the world around me. Thank you for showing me the power of the natural world. Thank you for giving me a sense of peace and tranquillity.

I promise to do my part to protect you and your many wonders. I will teach my children about the importance of conservation and sustainability. I will work to make the world a better place for all living things.

Thank you for everything, Nature.

With love,

N.B Singh

Contents

Preface

Quantum Mechanics is a fascinating and fundamental branch of physics that governs the behavior of matter and energy at the smallest scales. As students and researchers dive into the intricate world of quantum phenomena, the need for a quick and comprehensive reference becomes imperative.

Speed Quantum Mechanics: A Formula Handbook is crafted with the intention of providing a concise, formula-centric guide to quantum mechanics. This handbook is designed for those who seek rapid access to key equations, principles, and concepts in quantum physics. Whether you are a student preparing for exams, a researcher in need of a quick reference, or anyone interested in the beauty of quantum mechanics, this handbook aims to be your go-to resource.

Features of the Handbook

- **Formula-Centric Approach:** This handbook adopts a speed formula approach, presenting quantum mechanics concepts through a lens focused on key mathematical formulas and equations.

- **Comprehensive Coverage:** Encompassing essential topics such as wave-particle duality, quantum states, operators, time evolution, and more, the handbook strives to cover a broad spectrum of quantum mechanics in a concise manner.

- **Mathematical Rigor:** Formulas are presented using mathematical environments such as `dmath` and `align`, ensuring clarity and ease of understanding.

- **Quick Memorization Tips:** Throughout the handbook, fast memorization tips are provided to enhance the reader's ability to recall and apply critical concepts swiftly.

How to Use This Handbook

This handbook is organized into sections based on key quantum mechanics topics. Each section presents formulas, equations, and relevant information in a structured manner. Feel free to navigate to specific sections based on your current study or research needs.

As with any comprehensive reference, it is recommended to use this handbook in conjunction with textbooks and lecture materials for a well-rounded understanding of quantum mechanics.

I hope *Speed Quantum Mechanics: A Formula Handbook* serves as a valuable tool in your journey through the captivating world of quantum mechanics.

Chapter 1

Introduction

1.1 Overview

In this section, we aim for rapid understanding of key quantum mechanics concepts. Let's explore formulas for quick memorization:

1.1.1 Wave-Particle Duality

The de Broglie wavelength equation connects the momentum (p) and the wavelength (λ) of a particle:

$$\lambda = \frac{h}{p} \tag{1.1}$$

where h is Planck's constant.

1.1.2 Quantum States

The Schrödinger equation describes the time evolution of a quantum state $\psi(x,t)$:

$$i\hbar\frac{\partial}{\partial t}\psi(x,t) = \hat{H}\psi(x,t) \tag{1.2}$$

where $\hbar$ is the reduced Planck's constant and $\hat{H}$ is the Hamiltonian operator.

1.1.3   Operators and Observables

The expectation value of an observable $\hat{A}$ in a quantum state is given by:

$$\langle A \rangle = \int \psi^* \hat{A} \psi \, dx \tag{1.3}$$

1.1.4   Time Evolution

The probability density $P(x)$ of finding a particle at position x is determined by the modulus squared of the wavefunction:

$$P(x) = |\psi(x,t)|^2 \tag{1.4}$$

Fast memorization tip: Visualize these formulas as interconnected puzzle pieces, forming a holistic understanding of quantum mechanics.

Chapter 2

Key Concepts

2.1 Wave-Particle Duality

Wave-particle duality is a core concept in quantum mechanics, capturing the dual nature of particles. Let's dive into the key formulas for rapid understanding:

2.1.1 De Broglie Wavelength

The de Broglie wavelength (λ) relates the momentum (p) and Planck's constant (h) of a particle:

$$\lambda = \frac{h}{p} \tag{2.1}$$

2.1.2 Energy-Frequency Relation

The energy (E) of a photon is linked to its frequency (ν) through the Planck-Einstein relation:

$$E = h\nu \tag{2.2}$$

2.1.3 Kinetic Energy and Wavelength

The de Broglie wavelength can be connected to a particle's kinetic energy (E) and mass (m):

$$\lambda = \frac{h}{\sqrt{2mE}} \tag{2.3}$$

Fast memorization tip: Think of de Broglie as "wavelength wizard," relating momentum, energy, and mass through these snappy formulas.

2.2 Quantum States

Quantum states are described by wavefunctions (ψ), which embody the probabilities of finding a particle in a certain state. Let's explore the fundamental formulas for speedy recall:

2.2.1 Wavefunction

The wavefunction, denoted by $\psi(x,t)$, characterizes the state of a quantum system. It satisfies the Schrödinger equation:

$$i\hbar\frac{\partial}{\partial t}\psi(x,t) = \hat{H}\psi(x,t) \tag{2.4}$$

where $\hbar$ is the reduced Planck's constant, $\hat{H}$ is the Hamiltonian operator, and $\partial/\partial t$ represents the time derivative.

2.2.2 Probability Density

The probability density ($P(x)$) of finding a particle at position x is given by the modulus squared of the wavefunction:

$$P(x) = |\psi(x,t)|^2 \tag{2.5}$$

Fast memorization tip: Picture the wavefunction as a "probability painter," guiding the likelihood of finding a particle at different positions over time.

2.3 Operators and Observables

In quantum mechanics, operators play a crucial role in describing physical observables. Let's delve into the formulas for rapid understanding:

2.3.1 Operators

Operators in quantum mechanics are represented by Hermitian matrices. The general form of an operator $\hat{A}$ is:

$$\hat{A} = \sum_n a_n \left| \psi_n \right\rangle \left\langle \psi_n \right| \tag{2.6}$$

where a_n are eigenvalues and $\left| \psi_n \right\rangle$ are corresponding eigenvectors.

2.3.2 Expectation Value

The expectation value of an observable $\hat{A}$ in a quantum state $\psi(x, t)$ is given by:

$$\langle A \rangle = \int \psi^* \hat{A} \psi \, dx \tag{2.7}$$

2.3.3 Commutation Relations

For two operators $\hat{A}$ and $\hat{B}$, the commutation relation is expressed as:

$$[\hat{A}, \hat{B}] = \hat{A}\hat{B} - \hat{B}\hat{A} \tag{2.8}$$

The commutator $[\hat{A}, \hat{B}]$ provides information about the compatibility of measuring $\hat{A}$ and $\hat{B}$ simultaneously.

Fast memorization tip: Think of operators as "quantum tools" revealing observable properties and their relationships.

2.4 Time Evolution

Understanding the time evolution of quantum states is fundamental in quantum mechanics. Let's explore the key formulas for quick memorization:

2.4.1 Schrödinger Equation

The time evolution of a quantum state $\psi(x, t)$ is described by the Schrödinger equation:

$$i\hbar \frac{\partial}{\partial t}\psi(x, t) = \hat{H}\psi(x, t) \tag{2.9}$$

where $\hbar$ is the reduced Planck's constant and $\hat{H}$ is the Hamiltonian operator.

2.4.2 Time-Independent Schrödinger Equation

In situations where the Hamiltonian operator does not explicitly depend on time, the Schrödinger equation simplifies to the time-independent form:

$$\hat{H}\psi(x) = E\psi(x) \tag{2.10}$$

where E is the energy eigenvalue.

2.4.3 Stationary States

For systems in stationary states, the wavefunction $\psi(x, t)$ can be expressed as a product of spatial and temporal components:

$$\psi(x, t) = \phi(x)e^{-iEt/\hbar} \tag{2.11}$$

where $\phi(x)$ is the spatial part, and E is the energy eigenvalue.

Fast memorization tip: Visualize the Schrödinger equation as a "time sculptor," shaping the evolution of quantum states over time.

Chapter 3

Fundamental Formulas

3.1 Schrödinger Equation

The Schrödinger equation is the cornerstone of quantum mechanics, describing the time evolution of quantum states. Let's explore the key formulas for fast memorization:

3.1.1 Time-Dependent Schrödinger Equation

The time-dependent Schrödinger equation for a quantum state $\psi(x,t)$ is given by:

$$i\hbar \frac{\partial}{\partial t} \psi(x,t) = \hat{H}\psi(x,t) \tag{3.1}$$

where $\hbar$ is the reduced Planck's constant, $\hat{H}$ is the Hamiltonian operator, and $\frac{\partial}{\partial t}$ denotes the partial derivative with respect to time.

3.1.2 Time-Independent Schrödinger Equation

For situations where the Hamiltonian operator does not explicitly depend on time, the time-independent Schrödinger equation simplifies to:

$$\hat{H}\psi(x) = E\psi(x) \tag{3.2}$$

where E is the energy eigenvalue associated with the state.

3.1.3 Stationary States

In stationary states, the time-dependent wavefunction $\psi(x, t)$ can be expressed as a product of spatial and temporal components:

$$\psi(x, t) = \phi(x)e^{-iEt/\hbar} \tag{3.3}$$

where $\phi(x)$ is the spatial part, and E is the energy eigenvalue.

Fast memorization tip: Think of the Schrödinger equation as a "quantum time sculptor," governing the evolution of states and revealing the energy landscape.

3.2 Heisenberg Uncertainty Principle

The Heisenberg Uncertainty Principle is a fundamental concept in quantum mechanics, highlighting the inherent limits in simultaneously measuring certain pairs of properties. Let's explore the key formulas for fast memorization:

3.2.1 General Form

The Heisenberg Uncertainty Principle is mathematically expressed as:

$$\Delta x \cdot \Delta p \geq \frac{\hbar}{2} \tag{3.4}$$

where Δx is the uncertainty in position, Δp is the uncertainty in momentum, and $\hbar$ is the reduced Planck's constant.

3.2.2 Position-Momentum Form

The uncertainties in position and momentum are defined as:

$$\Delta x = \sqrt{\langle x^2 \rangle - \langle x \rangle^2} \tag{3.5}$$

$$\Delta p = \sqrt{\langle p^2 \rangle - \langle p \rangle^2} \tag{3.6}$$

where $\langle x \rangle$ and $\langle p \rangle$ are the expectation values of position and momentum, respectively.

3.2.3 Time-Energy Form

The Heisenberg Uncertainty Principle can also be expressed in terms of time and energy uncertainties:

$$\Delta E \cdot \Delta t \geq \frac{\hbar}{2} \tag{3.7}$$

where ΔE is the uncertainty in energy, and Δt is the uncertainty in time.

Fast memorization tip: Visualize the Heisenberg Uncertainty Principle as a "quantum tightrope," showcasing the inherent trade-off between certain pairs of measurements.

3.3 Probability Density

Understanding probability density is crucial in quantum mechanics as it represents the likelihood of finding a particle at a specific position. Let's explore the key formulas for fast memorization:

3.3.1 Definition

The probability density $P(x)$ is defined as the modulus squared of the wavefunction $\psi(x, t)$:

$$P(x) = |\psi(x, t)|^2 \tag{3.8}$$

3.3.2 Normalization Condition

For a normalized wavefunction $\psi(x, t)$, the integral of the probability density over all positions is equal to 1:

$$\int_{-\infty}^{\infty} |\psi(x, t)|^2 \, dx = 1 \tag{3.9}$$

3.3.3   Expectation Value of Position

The expectation value of position $\langle x \rangle$ for a wavefunction $\psi(x, t)$ is given by:

$$\langle x \rangle = \int_{-\infty}^{\infty} x |\psi(x, t)|^2 \, dx \tag{3.10}$$

Fast memorization tip: Envision probability density as a "quantum spotlight," revealing the likelihood of finding a particle at different positions.

3.4   Eigenstates and Eigenvalues

Eigenstates and eigenvalues are fundamental concepts in quantum mechanics, representing the states and corresponding values for which certain observables are well-defined. Let's explore the key formulas for fast memorization:

3.4.1   Eigenvalue Equation

For an observable represented by the operator $\hat{A}$, the eigenvalue equation is given by:

$$\hat{A} |\psi\rangle = a |\psi\rangle \tag{3.11}$$

where a is the eigenvalue and $|\psi\rangle$ is the corresponding eigenstate.

3.4.2   Normalization Condition

Eigenstates are normalized, satisfying the condition:

$$\langle \psi | \psi \rangle = 1 \tag{3.12}$$

where $\langle \psi |$ is the bra vector corresponding to the eigenstate $|\psi\rangle$.

3.4.3   Completeness Relation

The completeness relation expresses the identity operator $\hat{I}$ in terms of the complete set of eigenstates $\{|\psi_n\rangle\}$:

$$\hat{I} = \sum_n |\psi_n\rangle \langle \psi_n| \tag{3.13}$$

Fast memorization tip: Think of eigenstates as "observable beacons," providing well-defined values for quantum properties.

Chapter 4

Quantum Mechanics in 3D

4.1 Angular Momentum

Angular momentum is a crucial concept in quantum mechanics, describing rotational motion. Let's explore the key formulas for fast memorization:

4.1.1 Angular Momentum Operator

The angular momentum operator $\hat{L}$ is defined in terms of position and momentum operators as:

$$\hat{L} = \hat{r} \times \hat{p} \tag{4.1}$$

where $\hat{r}$ is the position operator and $\hat{p}$ is the momentum operator.

4.1.2 Eigenvalue Equation

The eigenvalue equation for the angular momentum operator is given by:

$$\hat{L} |\psi\rangle = \hbar l(l + 1) |\psi\rangle \tag{4.2}$$

where l is the quantum number associated with angular momentum and $\hbar$ is the reduced Planck's constant.

4.1.3   Angular Momentum Components

The components of the angular momentum operator in Cartesian coordinates are:

$$\hat{L}_x = \hat{y}\hat{p}_z - \hat{z}\hat{p}_y \tag{4.3}$$

$$\hat{L}_y = \hat{z}\hat{p}_x - \hat{x}\hat{p}_z \tag{4.4}$$

$$\hat{L}_z = \hat{x}\hat{p}_y - \hat{y}\hat{p}_x \tag{4.5}$$

Fast memorization tip: Envision angular momentum as a "quantum spinner," capturing the rotational properties of particles.

4.2   Spherical Coordinates

Spherical coordinates provide a convenient way to describe positions in three-dimensional space. Let's explore the key formulas for fast memorization:

4.2.1   Conversion from Cartesian Coordinates

The conversion from Cartesian coordinates (x, y, z) to spherical coordinates (r, θ, ϕ) is given by:

$$r = \sqrt{x^2 + y^2 + z^2} \tag{4.6}$$

$$\theta = \arccos\left(\frac{z}{r}\right) \tag{4.7}$$

$$\phi = \arctan\left(\frac{y}{x}\right) \tag{4.8}$$

4.2.2   Spherical Unit Vectors

The unit vectors in spherical coordinates are defined as:

$$\hat{r} = \sin\theta\cos\phi\,\hat{i} + \sin\theta\sin\phi\,\hat{j} + \cos\theta\,\hat{k} \tag{4.9}$$

$$\hat{\theta} = \cos\theta\cos\phi\,\hat{i} + \cos\theta\sin\phi\,\hat{j} - \sin\theta\,\hat{k} \tag{4.10}$$

$$\hat{\phi} = -\sin\phi\,\hat{i} + \cos\phi\,\hat{j} \tag{4.11}$$

4.2.3 Volume Element

The volume element dV in spherical coordinates is given by:

$$dV = r^2 \sin\theta \, dr \, d\theta \, d\phi \tag{4.12}$$

Fast memorization tip: Imagine spherical coordinates as a "quantum GPS," pinpointing locations in a three-dimensional space.

Chapter 5

Special Topics

5.1 Quantum Entanglement

Quantum entanglement is a unique feature of quantum mechanics. Let's explore the key formulas for fast memorization:

5.1.1 Entangled State

An entangled state of two particles is often represented as:

$$|\Psi\rangle = \frac{1}{\sqrt{2}} \left(|0\rangle_A |1\rangle_B - |1\rangle_A |0\rangle_B \right) \tag{5.1}$$

where $|0\rangle$ and $|1\rangle$ are basis states, and the subscripts A and B denote different particles.

5.1.2 Bell State

One of the famous entangled states, a Bell state, is given by:

$$|\Phi^+\rangle = \frac{1}{\sqrt{2}} \left(|0\rangle_A |0\rangle_B + |1\rangle_A |1\rangle_B \right) \tag{5.2}$$

5.1.3 Entanglement Criterion

For a general two-particle state $|\Psi\rangle$, an entangled state is indicated if the state cannot be written as a product of individual states:

$$|\Psi\rangle \neq |\psi\rangle_A \otimes |\phi\rangle_B \tag{5.3}$$

Fast memorization tip: Envision entanglement as a "quantum dance," where the steps of one particle are intimately linked with the steps of another.

5.2 Quantum Superposition

Quantum superposition is a fundamental concept where a quantum system can exist in multiple states simultaneously. Let's explore the key formulas for fast memorization:

5.2.1 Superposition Principle

The superposition principle states that if $|\psi_1\rangle$ and $|\psi_2\rangle$ are solutions to the Schrödinger equation, then any linear combination of them is also a solution:

$$|\Psi\rangle = c_1 |\psi_1\rangle + c_2 |\psi_2\rangle \tag{5.4}$$

where c_1 and c_2 are complex coefficients, and $|\Psi\rangle$ is a superposition state.

5.2.2 Probability in Superposition

The probability of measuring the system in state $|\psi_i\rangle$, given a superposition state $|\Psi\rangle$, is given by the squared modulus of the corresponding coefficient:

$$P(|\psi_i\rangle) = |c_i|^2 \tag{5.5}$$

5.2.3 EPR Paradox

The Einstein-Podolsky-Rosen (EPR) paradox involves entangled particles in a superposition of states. For example, an entangled state can be represented as:

$$|\Psi\rangle = \frac{1}{\sqrt{2}} \left(|0\rangle_A |1\rangle_B - |1\rangle_A |0\rangle_B \right) \tag{5.6}$$

Fast memorization tip: Picture superposition as a "quantum orchestra," where different states play harmoniously together.

5.3 Quantum Tunneling

Quantum tunneling is a quantum mechanical phenomenon where particles can penetrate through energy barriers. Let's explore the key formulas for fast memorization:

5.3.1 Tunneling Probability

The probability of quantum tunneling through a potential barrier is given by the transmission coefficient, T:

$$T = e^{-2\alpha d} \tag{5.7}$$

where α is the tunneling coefficient and d is the width of the barrier.

5.3.2 Tunneling Current

In the context of quantum mechanics in solid-state physics, the tunneling current I through a barrier can be expressed as:

$$I = \frac{V}{R} \propto e^{-2\alpha d} \tag{5.8}$$

where V is the voltage across the barrier, R is the barrier resistance, and α and d have similar meanings as before.

5.3.3 WKB Approximation

The semiclassical WKB (Wentzel-Kramers-Brillouin) approximation provides an estimate of the tunneling probability using the wavefunction phase change:

$$T \approx e^{-\frac{2}{\hbar} \int_{x_1}^{x_2} \sqrt{2m(V(x)-E)}\, dx} \tag{5.9}$$

where m is the particle mass, $V(x)$ is the potential energy, E is the energy of the particle, and x_1 and x_2 are the classical turning points.

Fast memorization tip: Imagine quantum tunneling as a "particle ninja," effortlessly passing through barriers.

Chapter 6

Advanced Quantum Mechanics

6.1 Variational Methods

Variational methods are powerful techniques in quantum mechanics for approximating the ground state energy of a system. Let's explore the key formulas for fast memorization:

6.1.1 Variational Principle

The variational principle states that the energy expectation value of a trial wavefunction ψ_T is an upper bound to the true ground state energy E_0:

$$E_0 \leq \frac{\langle \psi_T | \hat{H} | \psi_T \rangle}{\langle \psi_T | \psi_T \rangle} \tag{6.1}$$

where $\hat{H}$ is the Hamiltonian operator.

6.1.2 Variational Parameter

In variational methods, a parameter α is introduced in the trial wavefunction to optimize the energy. For example, in a simple trial wavefunction $\psi_T(x; \alpha)$, α

is the variational parameter.

6.1.3 Rayleigh-Ritz Method

The Rayleigh-Ritz method is often employed in variational calculations to find an upper bound for the ground state energy:

$$E_0 \leq \frac{\langle \psi_T | \hat{H} | \psi_T \rangle}{\langle \psi_T | \psi_T \rangle} \tag{6.2}$$

6.1.4 Optimization Condition

To find the optimal variational parameter, the condition for the energy to be minimized is given by:

$$\frac{\partial}{\partial \alpha} \left(\frac{\langle \psi_T | \hat{H} | \psi_T \rangle}{\langle \psi_T | \psi_T \rangle} \right) = 0 \tag{6.3}$$

Fast memorization tip: Think of variational methods as a "quantum sculptor," shaping trial wavefunctions to approximate ground state energies.

6.2 Perturbation Theory

Perturbation theory is a powerful tool for analyzing systems where the Hamiltonian can be separated into a known unperturbed part and a small perturbation. Let's explore the key formulas for fast memorization:

6.2.1 Perturbed Hamiltonian

Consider a Hamiltonian $\hat{H}$ that can be separated into an unperturbed part $\hat{H}_0$ and a perturbation term $\hat{V}$:

$$\hat{H} = \hat{H}_0 + \hat{V} \tag{6.4}$$

6.2.2 Non-Degenerate Perturbation Theory

For non-degenerate perturbation theory, where the eigenstates of $\hat{H}_0$ are non-degenerate, the first-order correction to the energy is given by:

$$E_n^{(1)} = \langle \psi_n^{(0)} | \hat{V} | \psi_n^{(0)} \rangle \tag{6.5}$$

6.2.3 Degenerate Perturbation Theory

For degenerate perturbation theory, where the eigenstates of $\hat{H}_0$ are degenerate, the first-order correction to the energy is given by:

$$E_n^{(1)} = \sum_{m \neq n} \frac{|\langle \psi_m^{(0)} | \hat{V} | \psi_n^{(0)} \rangle|^2}{E_n^{(0)} - E_m^{(0)}} \tag{6.6}$$

6.2.4 Brueckner's Theorem

Brueckner's theorem provides an expression for the second-order correction to the energy in degenerate perturbation theory:

$$E_n^{(2)} = \sum_{m \neq n} \frac{|\langle \psi_m^{(0)} | \hat{V} | \psi_n^{(0)} \rangle|^2}{E_n^{(0)} - E_m^{(0)}} + \sum_{m \neq n} \sum_{l \neq n} \frac{\langle \psi_m^{(0)} | \hat{V} | \psi_l^{(0)} \rangle \langle \psi_l^{(0)} | \hat{V} | \psi_n^{(0)} \rangle}{(E_n^{(0)} - E_m^{(0)})(E_n^{(0)} - E_l^{(0)})} \tag{6.7}$$

Fast memorization tip: Visualize perturbation theory as a "quantum puzzle solver," systematically correcting energies with perturbations.

6.3 Time-Dependent Perturbation

Time-dependent perturbation theory is a powerful tool for describing systems subject to time-varying perturbations. Let's explore the key formulas for fast memorization:

6.3.1 Time-Dependent Perturbation Hamiltonian

Consider a time-dependent perturbation $\hat{V}(t)$ acting on a system described by the unperturbed Hamiltonian $\hat{H}_0$. The total Hamiltonian is given by:

$$\hat{H}(t) = \hat{H}_0 + \hat{V}(t) \tag{6.8}$$

6.3.2　Interaction Picture

In the interaction picture, states and operators evolve with the perturbation. The time evolution operator $\hat{U}_0(t, t_0)$ is given by:

$$\hat{U}_0(t, t_0) = \exp\left(-\frac{i}{\hbar} \int_{t_0}^{t} \hat{H}_0(\tau)\, d\tau\right) \tag{6.9}$$

6.3.3　Time-Dependent Schrödinger Equation

The time-dependent Schrödinger equation in the interaction picture is:

$$i\hbar \frac{d}{dt} \lvert \psi(t) \rangle_I = \hat{V}_I(t) \lvert \psi(t) \rangle_I \tag{6.10}$$

where $\hat{V}_I(t) = \hat{U}_0^\dagger(t, t_0) \hat{V}(t) \hat{U}_0(t, t_0)$ is the perturbation in the interaction picture.

6.3.4　Fermi's Golden Rule

For transitions between states in the presence of a time-dependent perturbation, Fermi's Golden Rule gives the transition probability per unit time:

$$W_{fi} = \frac{2\pi}{\hbar} \left| \langle \psi_f | \hat{V}(t) | \psi_i \rangle \right|^2 \delta(E_f - E_i) \tag{6.11}$$

where $\lvert \psi_i \rangle$ and $\lvert \psi_f \rangle$ are the initial and final states, respectively.

Fast memorization tip: Picture time-dependent perturbation as a "quantum conductor," orchestrating transitions between states with a time-varying influence.

Chapter 7

Relativistic Quantum Mechanics

7.1 Dirac Equation

The Dirac equation describes the behavior of relativistic electrons and other fermions. Let's explore the key formulas for fast memorization:

7.1.1 Dirac Equation in Natural Units

In natural units ($\hbar = c = 1$), the Dirac equation for a free particle is given by:

$$(i\gamma^\mu \partial_\mu - m)\psi = 0 \tag{7.1}$$

where ψ is the Dirac spinor, m is the mass of the particle, ∂_μ is the partial derivative with respect to spacetime coordinates, and γ^μ are the Dirac matrices.

7.1.2 Dirac Matrices

The Dirac matrices satisfy the anticommutation relation:

$$\{\gamma^\mu, \gamma^\nu\} = 2g^{\mu\nu} \tag{7.2}$$

where $\{\cdot, \cdot\}$ denotes the anticommutator, and $g^{\mu\nu}$ is the Minkowski metric.

7.1.3 Plane-Wave Solution

A plane-wave solution of the Dirac equation can be expressed as:

$$\psi(x) = u(p)e^{-ip\cdot x} + v(p)e^{ip\cdot x} \tag{7.3}$$

where $u(p)$ and $v(p)$ are spinors representing positive and negative energy solutions, respectively, and p is the four-momentum.

7.1.4 Charge Conjugation

The charge conjugation operation transforms a Dirac spinor to its antiparticle:

$$\psi^C = C\bar{\psi}^T \tag{7.4}$$

where C is the charge conjugation matrix, $\bar{\psi}$ is the Dirac adjoint, and T denotes the transpose.

Fast memorization tip: Envision the Dirac equation as a "quantum relativist," describing the dance of relativistic particles in spacetime.

7.2 Spinors

Spinors are mathematical objects used to describe the intrinsic angular momentum (spin) of particles. Let's explore the key formulas for fast memorization:

7.2.1 Dirac Spinor

The Dirac spinor is a four-component complex vector describing the state of a spin-1/2 particle. In natural units, it is denoted as:

$$\psi = \begin{pmatrix} \psi_1 \\ \psi_2 \\ \psi_3 \\ \psi_4 \end{pmatrix} \tag{7.5}$$

7.2.2 Pauli Matrices

The Pauli matrices, denoted as σ^i for $i = 1, 2, 3$, are 2x2 matrices representing the spin operators. They satisfy the anticommutation relation:

$$\{\sigma^i, \sigma^j\} = 2\delta^{ij}\mathbf{I} \tag{7.6}$$

where δ^{ij} is the Kronecker delta and $\mathbf{I}$ is the 2x2 identity matrix.

7.2.3 Spinor Transformation

Under a Lorentz transformation, the Dirac spinor transforms as:

$$\psi' = S(\Lambda)\psi \tag{7.7}$$

where $S(\Lambda)$ is the spinor transformation matrix associated with the Lorentz transformation Λ.

7.2.4 Weyl Spinors

Weyl spinors are two-component spinors representing left- or right-handed chiral states. In natural units, a Weyl spinor is denoted as:

$$\chi_L = \begin{pmatrix} \chi_1 \\ \chi_2 \end{pmatrix}, \quad \chi_R = \begin{pmatrix} \chi_3 \\ \chi_4 \end{pmatrix} \tag{7.8}$$

Fast memorization tip: Picture spinors as "quantum dancers," capturing the intrinsic dance of particles in the quantum world.

Chapter 8

Quantum Information

8.1 Qubits and Quantum Computing

Qubits are the fundamental units of information in quantum computing. Let's explore the key formulas for fast memorization:

8.1.1 Qubit Representation

A qubit is a quantum two-level system represented by a state vector:

$$|\psi\rangle = \alpha |0\rangle + \beta |1\rangle \tag{8.1}$$

where α and β are complex probability amplitudes, and $|0\rangle$ and $|1\rangle$ are the computational basis states.

8.1.2 Quantum Gates

Quantum gates perform operations on qubits. For example, the Hadamard gate acts on a single qubit as:

$$H |0\rangle = \frac{1}{\sqrt{2}}(|0\rangle + |1\rangle) \tag{8.2}$$

$$H |1\rangle = \frac{1}{\sqrt{2}}(|0\rangle - |1\rangle) \tag{8.3}$$

8.1.3 Quantum Entanglement in Bell State

The Bell state, representing entanglement between two qubits, is given by:

$$|\Phi^+\rangle = \frac{1}{\sqrt{2}}(|00\rangle + |11\rangle) \tag{8.4}$$

8.1.4 Quantum Superposition and Measurement

A qubit in superposition can be expressed as:

$$|\psi\rangle = \frac{1}{\sqrt{2}}(|0\rangle + |1\rangle) \tag{8.5}$$

Upon measurement, the qubit collapses to one of the basis states with probabilities $|\alpha|^2$ and $|\beta|^2$.

Fast memorization tip: Visualize qubits as "quantum acrobats," performing extraordinary feats of information processing in superposition and entanglement.

8.2 Quantum Cryptography

Quantum cryptography leverages the principles of quantum mechanics for secure communication. Let's explore the key formulas for fast memorization:

8.2.1 Quantum Key Distribution (QKD)

QKD is a fundamental protocol in quantum cryptography. The BBM92 protocol involves the following steps:

Step 1: Preparation   Alice prepares a qubit in one of the states: (8.6)

$$|0\rangle , |1\rangle , |+\rangle = \frac{1}{\sqrt{2}}(|0\rangle + |1\rangle), |-\rangle = \frac{1}{\sqrt{2}}(|0\rangle - |1\rangle)$$

Step 2: Transmission   Alice sends the qubits to Bob over a quantum channel.

Step 3: Measurement   Bob randomly measures the qubits in either the standard basis ($|0\rangle$ and $|1\rangle$) or the Hadamard basis ($|+\rangle$ and $|-\rangle$).

Step 4: Key Extraction   Alice and Bob publicly announce their measurement bases for each qubit. They keep the bits measured in the same basis,forming the raw key.

Step 5: Information Reconciliation and Privacy Amplification

Alice and Bob perform error correction (information reconciliation) and distill a shorter but secure key (privacy amplification).

8.2.2   BB84 Protocol - Basis Rotation

The BB84 protocol introduces basis rotation in qubit transmission. Alice and Bob use the standard ($|0\rangle$ and $|1\rangle$) and diagonal ($|+\rangle$ and $|-\rangle$) bases, creating a more robust key distribution.

$$\text{Standard Basis:}\quad |0\rangle , |1\rangle$$
$$\text{Diagonal Basis:}\quad |+\rangle = \frac{1}{\sqrt{2}}(|0\rangle + |1\rangle), |-\rangle = \frac{1}{\sqrt{2}}(|0\rangle - |1\rangle)$$

Fast memorization tip: Envision quantum cryptography as a "quantum lock and key," where secure communication relies on the principles of quantum mechanics.

Formula Quick Reference

8.3 Common Constants

Here are some commonly used mathematical constants:

8.3.1 Euler's Number (e)

Euler's number is an important mathematical constant and the base of the
natural logarithm:

$$e \approx 2.71828 \tag{8.7}$$

8.3.2 Pi (π)

Pi is the ratio of the circumference of a circle to its diameter:

$$\pi \approx 3.14159 \tag{8.8}$$

8.3.3 Golden Ratio (ϕ)

The golden ratio is a special irrational number that appears in various mathematical and artistic contexts:

$$\phi = \frac{1 + \sqrt{5}}{2} \approx 1.61803 \tag{8.9}$$

8.3.4 Imaginary Unit (i)

The imaginary unit is defined as the square root of -1:

$$i^2 = -1 \tag{8.10}$$

8.3.5 Avogadro's Number (N_A)

Avogadro's number represents the number of atoms or molecules in one mole of a substance:

$$N_A \approx 6.022 \times 10^{23}\,\text{mol}^{-1} \tag{8.11}$$

8.3.6 Speed of Light in a Vacuum (c)

The speed of light is a fundamental constant in physics:

$$c \approx 3.00 \times 10^8\,\text{m/s} \tag{8.12}$$

8.3.7 Planck's Constant (h)

Planck's constant relates the energy of a photon to its frequency:

$$h \approx 6.626 \times 10^{-34}\,\text{J} \cdot \text{s} \tag{8.13}$$

8.3.8 Gravitational Constant (G)

The gravitational constant appears in the law of universal gravitation:

$$G \approx 6.674 \times 10^{-11}\,\text{m}^3\,\text{kg}^{-1}\,\text{s}^{-2} \tag{8.14}$$

Fast memorization tip: Remember these constants as the "essential tools" for various mathematical and scientific calculations.

8.4 Conversion Factors

Here are some commonly used conversion factors:

8.4.1 Length

Metric to Imperial

$$1\,\text{meter} \approx 3.2808\,\text{feet} \tag{8.15}$$

$$1\,\text{centimeter} \approx 0.0328\,\text{feet} \tag{8.16}$$

Imperial to Metric

$$1\,\text{foot} \approx 0.3048\,\text{meters} \tag{8.17}$$

$$1\,\text{inch} \approx 0.0254\,\text{meters} \tag{8.18}$$

8.4.2 Mass

Metric to Imperial

$$1\,\text{kilogram} \approx 2.2046\,\text{pounds} \tag{8.19}$$

$$1\,\text{gram} \approx 0.0353\,\text{ounces} \tag{8.20}$$

Imperial to Metric

$$1\,\text{pound} \approx 0.4536\,\text{kilograms} \tag{8.21}$$

$$1\,\text{ounce} \approx 28.3495\,\text{grams} \tag{8.22}$$

8.4.3 Volume

Metric to Imperial

$$1\,\text{liter} \approx 0.2642\,\text{gallons} \tag{8.23}$$

$$1\,\text{milliliter} \approx 0.001\,\text{liters} \tag{8.24}$$

Imperial to Metric

$$1\,\text{gallon} \approx 3.7854\,\text{liters} \tag{8.25}$$

$$1\,\text{fluid ounce} \approx 0.0296\,\text{liters} \tag{8.26}$$

8.4.4   Temperature

$$^\circ\text{Celsius} \approx \left(\frac{^\circ\text{Fahrenheit} - 32}{1.8} \right) \tag{8.27}$$

$$^\circ\text{Fahrenheit} \approx (1.8 \times\,^\circ\text{Celsius} + 32) \tag{8.28}$$

Fast memorization tip: Think of conversion factors as "bridges" connecting different measurement systems, making transitions smooth.